Unraveling Existence

A Journey Through Human Lifespan

Freudian Trips

Copyright Page

Published by Omniterra Media Inc

First Edition

Visit the author's website at www.freudiantrips.com

Disclaimer

The views and opinions expressed in this book are those of the author(s) and do not necessarily reflect the official policy or position of any other agency, organization, employer, or company. The contents of this book are for informational and educational purposes only and are not intended to serve as professional advice, diagnosis, or treatment.

The information provided in this book is believed to be accurate and reliable as of the date of publication. However, it may include some errors or inaccuracies, and no warranty or guarantee is provided regarding the accuracy, timeliness, or applicability of the content.

Readers are encouraged to consult with professional philosophers, educators, or other qualified professionals where appropriate for personalized advice. The author(s) and publisher shall not be liable for any loss, damage, or harm caused or alleged to be caused, directly or indirectly, by the information or ideas contained, suggested, or referenced in this book.

By reading this book, the reader acknowledges and agrees that they are solely responsible for how they interpret and apply the information contained herein.

This book may also include references to other works, studies, and sources. These references are provided for further reading and exploration and do not imply endorsement or validation of the specific theories, viewpoints, or interpretations presented in those works.

Chapter 1: An Awakening

The Tapestry of Life

Imagine you're standing in front of a grand tapestry, woven with threads of various colors, textures, and patterns. Each thread represents a moment, an experience, or a phase of a human's life. As you move closer, the minute details become clearer, and you notice that each thread has its own story. This intricate tapestry, my dear reader, is a metaphor for human development, a fascinating journey from the cradle to the grave.

A Journey Like No Other

Human development isn't just about the physical changes that turn a tiny baby into an elderly person. It's also about the emotional roller coasters, the forging of friendships, the dreams that keep us up at night, and the wisdom we gather along the way.

You see, understanding our journey, both its ups and downs, is akin to reading a map. Just as a map helps us navigate unfamiliar terrains, understanding the stages of human life helps us better comprehend ourselves and our loved ones.

Why Should You Care?

You might ask, "Why does any of this matter to me? I've lived a good chunk of my life already." Or perhaps, "Isn't this something for psychologists or educators to worry about?" Well, the truth is, no matter where you are in your life's journey, having insight into these phases can be a powerful tool.

For Reflection: By understanding where you've been, you can make sense of your past, and you might even view certain events in a new light.

For Relationships: Knowing what a child, parent, partner, or friend might be going through can help foster empathy and strengthen bonds.

For Anticipation: It can prepare you for the road ahead. Just like knowing there's a bend in the road lets you steer safely, understanding future phases of life can help you navigate them better.

A Teaser of What's to Come

The coming chapters will unravel this journey, taking you through every stage, every twist, and every turn. You'll find stories, real-world examples, and simple explanations that aim to enlighten, entertain, and perhaps even surprise you.

As we move forward, keep the tapestry in mind. Remember that every stage, no matter how challenging, adds a unique thread to the fabric of our existence. By the end, we hope you'll have a deeper appreciation for the intricate beauty of life, both in its individual threads and its collective majesty.

And with that, dear reader, let's set sail into the vast ocean of human development, where each wave and ripple holds a story waiting to be told. Welcome to the awakening.

Chapter 2: Prenatal Waves: The Dawn of Life

Before the First Hello

Imagine a secret world, hidden deep within the embrace of a mother's body. It's a place of magic where a tiny spark grows into a being with a heartbeat, fingers, and even the beginnings of dreams. This mystical realm? It's the womb, and the adventure we're about to embark on is the incredible journey of prenatal development.

Nature's Symphony in Three Movements

Like a beautifully composed symphony, the growth inside the womb takes place in three main movements or stages. Let's dive into each one, understanding its unique rhythm and melody.

The Opening Note: The Zygote Stage

This is where it all begins – with a single cell! When the mother's egg and the father's sperm unite, they form what's called a zygote. It might be hard to believe, but this tiny cell carries the blueprint for every detail of a person's life — from the color of their eyes to their love for chocolate (or lack of it). For about a week, this zygote journeys to find its home in the mother's womb and starts to divide and grow.

The Crescendo: The Embryo Stage

From the second week to the end of the second month, our little cell cluster takes on a new name: the embryo. This is a bustling period! The heart begins its steady drumbeat, and other vital organs like the brain and spine start to form. Even tiny buds that will become arms and legs sprout. It's a stage of rapid transformation, where the essential systems of the body are laid down.

The Grand Finale: The Fetus Stage

From the third month till birth, the embryo graduates to being called a fetus. If you could peek inside the womb now, you'd recognize it as a little human. It stretches, yawns, and might even suck its thumb! Over these months, it will grow bigger and stronger, preparing for its grand debut into the outside world.

The Ocean's Influence on the Shore

Now, you may wonder, "Why is this process important for me to understand?" Well, think of the womb as an ocean and the growing baby as a shore being shaped by its waves. The environment inside the womb can influence how the baby develops. For instance, the foods a mother eats, her emotions, and even the sounds from the outside world can play a part in shaping this new life. These influences can have long-term effects on the child's health, temperament, and even their likes and dislikes.

A Marvel to Behold

The prenatal journey is nothing short of a miracle. In just nine months, a single cell evolves into a living, breathing human being, complete with thoughts, feelings, and dreams. Understanding this journey helps us appreciate the complexity and beauty of life. It also underscores the significance of care, love, and the environment during these crucial months.

As we close this chapter, take a moment to marvel at the wonder of life. How, in a hidden sanctuary, a unique blend of nature and nurture crafts the story of an individual even before they step into the light of day. The dawn of life, as you've seen, is truly a magical symphony.

Chapter 3: The Uncharted Land: Birth to Toddlerhood

First Steps in a Vast World

Imagine setting foot on an unexplored planet. Every sight is new, every sound an enchantment, and every touch a revelation. This is the world of a newborn — an endless landscape of wonder waiting to be discovered.

From their first cry to their first wobbly steps, the journey from birth to toddlerhood is a whirlwind adventure of growth and learning. Let's walk alongside these tiny explorers and witness their astonishing milestones.

The Physical Adventure: Building Blocks of Movement

The Grand Entrance: Birth

Our adventure begins with the drama of birth. From a cozy, watery world, our little hero emerges, taking their first breath of fresh air. It's their debut on the stage of life.

Learning to Control: 0-3 months

At first, babies might seem like they're just lying around, but there's a lot going on. They're starting to hold their heads up, kick their little feet, and even reach out and grasp things (like your finger!).

Rolling and Sitting: 4-7 months

Soon, lying down isn't enough! They roll over, a mini-acrobatic feat. And then, with some effort, they start sitting up, observing the world from a brand-new perspective.

On the Move: 8-12 months

The world becomes a playground as babies start to crawl, pulling themselves up, and taking those monumental first steps.

The Cognitive Quest: Discovering and Thinking

Seeing and Recognizing: 0-3 months

A baby's vision starts blurry but sharpens quickly. They begin to recognize familiar faces, especially the comforting sight of their caregivers.

Exploring with Hands and Mouth: 4-7 months

Everything, and we mean everything, becomes a tool for learning. Whether it's a toy, a spoon, or their own toes, babies will grab and often try to eat it. It's their way of understanding the world.

Problem Solving: 8-12 months

Babies begin to show early problem-solving skills. They understand that things exist even if they can't see them (peek-a-boo becomes a favorite game) and will look for a toy that's been hidden.

The Social-Emotional Voyage: Feeling and Connecting

Bonding and Attachment: 0-3 months

Babies quickly develop deep bonds with their main caregivers. They communicate not with words, but with cries, coos, and smiles.

Discovering Emotions: 4-7 months

Laughter, joy, curiosity, frustration— babies start displaying a rainbow of emotions. They recognize emotions in others, too.

Claiming Independence: 8-12 months

It's not a teen rebellion, but babies begin to assert their independence. They might have favorite toys, foods, and even show some early signs of temperament.

This uncharted land of birth to toddlerhood is a time of rapid, remarkable change. With each day, our tiny explorers gain skills, understanding, and a deeper connection to the world around them.

As we close this chapter, let's marvel at the resilience and determination of these little ones. Their journey is a reminder that every giant leap begins with a small, uncertain step — and every step, no matter how wobbly, is worth celebrating.

Chapter 4: Ebb and Flow: The Preschool Years

Setting Sail on New Adventures

Imagine standing at the edge of the sea, toes sinking into wet sand, eyes fixed on the horizon. Waves rush in, each one bringing new treasures and surprises. This is the essence of the preschool years – a time when the world, like the ocean, is both vast and inviting, filled with endless wonders.

Let's set sail on this journey, navigating through the tidal currents of language, play, friendships, and the nurturing harbors of early learning environments.

Language: From Babble to Stories

Evolving Expressions: Gone are the days of mere coos and cries. Now, our little adventurers are experimenting with words, forming sentences, and voicing their thoughts, sometimes with amusing results! "I big boy now!" might be a proud declaration you hear.

Question Flood: "Why is the sky blue?" "Where do babies come from?" The preschool years are inundated with questions. It's not just about curiosity; it's their way of understanding the world around them.

Storytime Magic: Stories play a pivotal role during this time. Whether it's a fairy tale, a personal anecdote, or a made-up yarn, stories help kids understand emotions, cultures, and values.

Play: More Than Just Fun

Imagination's Playground: Give a child a cardboard box, and it can turn into a spaceship, a castle, or a dragon's lair. Through imaginative play, children learn creativity, problem-solving, and empathy.

Mimicking the World: Watch a preschooler play, and you might see them imitating adults – cooking, driving, or even "going to work". This mimicry helps them understand different roles in society.

Learning Rules and Taking Turns: Group games introduce kids to rules, teamwork, and the idea that sometimes, you have to wait your turn.

Friendships: The First Tastes of Social Life

Bonding over Toys: Initially, play might be solitary or parallel (playing next to but not with others). Soon, shared interests like a favorite toy or game become the foundation of budding friendships.

Ups and Downs: These early friendships come with their set of squabbles – "That's MY toy!" But with guidance, kids start learning about sharing, empathy, and resolving conflicts.

Expressing Emotions: Preschoolers are bundles of emotions. They begin to learn the words for their feelings and how to express them, whether it's the frustration of a toppled block tower or the joy of a new discovery.

The Haven of Early Learning Environments

Structured Play: Preschools and early learning centers offer structured play – activities designed to be fun yet educational, helping children hone their cognitive and motor skills.

Socialization: These environments are mini-societies where children learn about diversity, cooperation, and the joy of shared experiences.

The Value of Guidance: Trained educators play a crucial role in guiding children, ensuring they're not just playing and learning, but also feeling safe and valued.

The ebb and flow of the preschool years is a dance of growth, exploration, and self-discovery. These early experiences, like the waves shaping the shore, leave a lasting imprint, molding our young ones into curious learners and empathetic individuals.

In the gentle push and pull of these years, we find the foundation of lifelong learning and the beginning of many wonderful voyages to come. So, as we drop anchor here, remember that every ocean, no matter how vast, starts with a single drop – just like every great journey begins in these tender, transformative years.

Chapter 5: The Turbulent Tides: The School-Age Child

Embarking on a Vibrant Voyage

Imagine being on a boat in the middle of the sea. At times, the waters are calm, reflecting the serene blue of the sky. But occasionally, the waves swell and crash, challenging the boat's course. This is the journey of a school-age child, a mix of tranquil discoveries and stormy challenges, all against the backdrop of the vast ocean called 'school'.

Join me as we navigate the ever-changing waters of middle childhood, taking in the sights and sounds of cognitive, physical, and socio-emotional horizons.

Cognitive Capers: The Joy of Knowing and Growing

Eager Explorers: As children venture into school, their thirst for knowledge becomes evident. Math problems, science experiments, or history tales, each subject broadens their understanding of the world.

Critical Thinkers: No longer accepting everything at face value, they begin to question, analyze, and form their own opinions. "Why did this historical event happen?" "How does this experiment work?"

Memory Masters: Remembering multiplication tables, the steps of a dance, or even lines in a school play. This age witnesses an expansion in memory capabilities.

Physical Pursuits: Active and Agile

Growing Pains and Gains: Those pants from last year might feel a tad short. Rapid growth spurts are common, often accompanied by an increase in appetite.

Masters of Movement: Climbing trees, mastering the bicycle, or trying out for the school's soccer team. Their improved motor skills let them explore a plethora of physical activities.

Health Habits: It's not just about growth; it's about understanding health. Kids start realizing the importance of nutrition, exercise, and even mental well-being.

Socio-emotional Seas: Finding Their Place

The School Ship: School isn't just a place of learning. It's where children find their first sense of identity outside the home. They're no longer just "Jamie's son" but also "the fastest runner" or "the best at math."

Navigating Friendships: Friendships take on new dimensions. There are best friends, groups, and unfortunately, sometimes, cliques. Loyalties are tested, secrets are shared, and bonds, both fragile and strong, are formed.

Riding the Emotional Waves: This age is a rollercoaster of emotions. Elation at an 'A' grade can quickly shift to dejection from a playground disagreement. They're learning to manage these emotions, seeking both independence and validation.

The Peer Port: An Anchor and Challenge

School-age children increasingly look towards their peers for cues. "What's cool?" "How should I behave in this situation?" While peers offer support and camaraderie, there's also the challenge of peer pressure. Learning to stay true to oneself while fitting in becomes a pivotal lesson.

As we dock at the end of this chapter, let's take a moment to appreciate the resilience of our young navigators. The school-age journey, with its ebbs and flows, shapes their character, intellect, and heart. And as any seasoned sailor would attest, it's not the calm but the storm that makes a skilled sailor. In these turbulent tides, our school-age children are not just surviving; they are thriving, learning, and evolving.

Chapter 6: The Rough Seas: Adolescence

Journeying through Tempests and Tranquility

Imagine steering a ship through unpredictable waters, with serene stretches suddenly giving way to tempestuous storms, with uncharted islands and captivating horizons awaiting discovery. This is adolescence: a tumultuous yet transformative journey from childhood's shores to the vast ocean of adulthood.

Join me as we sail through this rich tapestry of change, unraveling the mysteries of puberty, the quests for identity, and the dynamic landscapes of young minds and hearts.

The Biological Tide: Puberty's Puzzles

Awakening of the Body: As teens bid farewell to their childlike frames, they witness the onset of puberty. Boys hear their voices crack and deepen, while girls might notice their first period. It's the body's way of saying, "You're growing up!"

Growth Spurts and Giggles: Ever notice a teen outgrow their shoes in what feels like mere months? Rapid growth spurts are common. So is a mix of clumsiness and grace, as they get accustomed to their evolving physique.

Brain Waves: Not only is the body evolving, but so is the brain! The teenage brain changes significantly, especially in regions related to impulse control and decision-making.

The Psychological Voyage: In Search of the Self

Who Am I?: Teens often embark on a profound quest for identity. They may experiment with different styles, hobbies, and even ideologies, trying to figure out where they fit in the grand mosaic of life.

Emotional Rollercoasters: Hormonal changes, combined with life's pressures, can make emotions intense and sometimes overwhelming. One moment they're on top of the world, and the next, they might seek solitude to nurse a heartache.

Dreams and Aspirations: Adolescents begin to look towards the future. They ponder careers, their roles in society, and their personal aspirations, sketching out blueprints of their dreams.

The Social Spectrum: Bonds and Boundaries

The Peer Compass: Friends are no longer just playmates; they become confidants, role models, and sometimes, the yardstick by which one measures oneself. The opinions and validation of their peer group can deeply influence an adolescent's choices.

Romantic Currents: Heartbeats might race with that first crush or the shy blush of a first date. Romantic relationships introduce teens to a new realm of feelings, challenges, and joys.

Parents: The Anchors and Adversaries: As teens strive for independence, they might sometimes see parents as adversaries. But beneath the surface, the parental anchor remains vital, offering security and guidance amid the turbulent seas.

Navigating adolescence is like mastering the art of sailing through both calm waters and raging storms. It's a time of exploration, introspection, trials, and triumphs. As we moor our ship at the end of this chapter, it's crucial to understand and respect the depth and complexity of this phase.

Adolescence isn't just about growing pains; it's about finding oneself, forging connections, and shaping the future. As with any voyage, there are challenges, but with the right guidance and understanding, these rough seas can lead to the most exhilarating adventures and discoveries.

Chapter 7: The Ocean's Depths: Early Adulthood

Diving into Uncharted Waters

Imagine standing on a ship's deck, gazing at the vast expanse of the ocean. Beneath the shimmering surface lie hidden depths, treasures, and mysteries waiting to be explored. This is early adulthood: an exhilarating plunge into the deep sea of responsibilities, choices, dreams, and legacies.

Join me on this dive as we journey through the challenges and charms of career paths, budding families, and the navigation of life's most profound decisions.

Mapping the Career Currents

Setting Sail: For many, early adulthood marks the transition from academia to the professional realm. It's about finding one's footing in the corporate world, the arts, or any other chosen field.

Charting the Course: The initial years can be a mix of excitement and anxiety. Am I on the right path? Should I switch careers? Pursue further education? These questions are the crossroads of our professional journey.

The Quest for Balance: As work demands increase, young adults often grapple with balancing their job roles with personal passions and relaxation. It's a dance between ambition and well-being.

The Voyage of Relationships and Families

New Horizons: Relationships take on deeper dimensions. It might be about finding a life partner, deepening bonds, or navigating the complexities of long-term commitment.

The Adventure of Parenthood: For some, early adulthood ushers in the joys and challenges of parenthood. Sleepless nights, baby's first steps, and the profound realization of being responsible for another life.

Family Dynamics: It's not just about the nuclear family. Relationships with parents, now as an adult, and connections with siblings evolve. There's a blending of the roles of being cared for and becoming a caregiver.

Choices and Tides: The Crux of Adulthood

Decision Depths: Every day seems to bring crucial decisions. Buying a home, managing finances, planning for the future - these choices mold the trajectory of our adult lives.

The Responsibility Reef: With freedom comes responsibility. Whether it's paying bills, caring for a family member, or fulfilling work duties, early adulthood is often about juggling varied roles.

Self-Reflection and Growth: Amidst the whirlwind of responsibilities, introspection becomes essential. Who have I become? Am I true to my values? This self-reflection fuels personal growth.

Emerging from the depths of early adulthood, one gains a treasure trove of experiences, memories, and insights. It's a phase of life that's as challenging as it is rewarding, and each decision, each milestone, adds a unique shade to the canvas of life.

While early adulthood feels like deep-sea diving, with its pressures and profundities, it's also where we often find our most precious pearls of wisdom, love, and purpose. As we anchor this chapter, remember: the ocean's depths are where the true wonders lie, waiting for the brave and the curious to discover and cherish.

Chapter 8: Navigating the High Seas: Middle Adulthood

Steering through Waves and Wonders

Picture yourself as the seasoned captain of a ship, with the confidence of many voyages behind you, yet with uncharted waters still ahead. This ship, seasoned by time and tide, is middle adulthood. It's an era of introspection, evolution, and the embrace of both life's tempests and treasures.

Embark with me on this chapter's journey, where we'll explore the ebbs and flows of health, cognition, professional aspirations, and the intricate dance of relationships during these pivotal years.

The Winds of Physical Change

Weathered by Time: As the years accumulate, our bodies naturally undergo changes. Some may notice a silver strand in their hair, others might experience the creak of a joint or two. But with these signs of aging come resilience and endurance forged by time.

Health Horizons: Regular health check-ups become essential. With a proactive approach, many find this phase to be one of vitality and wellness, celebrating the body's journey and caring for it with renewed intent.

Fitness and Fulfillment: Many rediscover their passion for physical activity, be it through yoga, dancing, hiking, or any sport that resonates. It's about celebrating mobility and embracing holistic well-being.

Cognitive Currents: The Depth and Breadth of Thought

Wisdom's Whisper: While certain cognitive functions might slow down a tad, middle adulthood often ushers in enhanced problem-solving skills, judgment, and, most importantly, wisdom gleaned from life's myriad experiences.

Continual Learning: The mind, ever-curious, delves into new interests. Some might take up a new language, others could dive into philosophy or the arts. It's a testament to the adage: You're never too old to learn.

Memory's Melody: Recollections gain richness, with memories serving as both lessons from the past and guides for the future.

The Career Compass: Mastery and Meaning

Climbing or Cruising: For some, middle adulthood is about reaching the zenith of their careers, while others might seek a more relaxed, fulfilling path, prioritizing passion over pace.

Legacy and Leadership: Many step into mentorship roles, guiding younger colleagues and reflecting on the legacy they aim to leave in their professional sphere.

Possibilities of Pivoting: It's also a time when some might consider career shifts, following latent passions or dreams previously set aside.

Harbors of the Heart: Relationships and Revelations

Deepening Bonds: Relationships, like fine wine, often deepen with time. Partners navigate the challenges together, cherishing the shared journey.

Empty Nest or Bursting Roost: As children grow and perhaps leave home, couples rediscover each other, while some homes buzz with the joy of grandchildren.

Friendships and Foundations: Lifelong friendships are treasures, offering comfort and camaraderie. New friendships too blossom, based on shared interests and life phases.

Middle adulthood, with its vast oceanic expanse, presents challenges but also unparalleled beauty. It's a journey of reflection, re-evaluation, and reinvigoration. As we close this chapter, take a moment to appreciate the captain within you, navigating life's high seas with grace, grit, and an ever-glowing spirit of adventure.

Chapter 9: The Silver Shores: Late Adulthood

The Twilight's Golden Glow

Envision a serene beach at twilight, with waves gently caressing the silver sands, the horizon painted with hues of gold and lavender. This tranquil setting mirrors late adulthood: a phase imbued with reflections, revelations, and the richness of a life well-lived.

Let's stroll together along these silver shores, exploring the landscapes of cognition, health, evolving social roles, and the profound contemplation of life's ephemeral nature.

Cognition: The Tides of Thought

Memory's Mosaic: Over the years, our brain has woven a vast tapestry of memories. While short-term recall might sometimes be elusive, the past often presents itself in vivid detail, offering both nostalgia and wisdom.

Wisdom's Wellspring: The depth of understanding and perspective that late adulthood brings is incomparable. Every experience, every lesson learned becomes a beacon for oneself and others.

Cherishing Challenges: Some cognitive tasks might take a tad longer, but the joy of solving puzzles, reading, or engaging in thoughtful discussions remains undiminished.

Health: The Dance of Delicacy and Determination

Navigating Nuances: The body, in its seasoned state, may have its set of challenges. Yet, with proactive care, regular check-ups, and perhaps a few lifestyle adjustments, many find a rhythm that celebrates life's dance.

Holistic Harmony: Activities like meditation, gentle yoga, or even daily walks become cherished routines. They're not just about physical well-being, but soulful serenity.

Gratitude and Grace: Every sunrise, every laughter-filled moment becomes a testament to life's beauty, a reason to cherish and celebrate.

Social Roles: Evolving Echoes

The Mentor's Mantle: With a reservoir of experiences, many in late adulthood naturally slip into roles of mentorship, offering guidance, stories, and invaluable insights to younger generations.

Communities and Connections: Social circles might evolve. Some relationships deepen, while new friendships, often with peers sharing similar life phases, offer comfort and camaraderie.

Roles Reimagined: From active professionals to retirees, from parents to doting grandparents, the shift in roles brings with it new adventures, responsibilities, and joys.

Mortality: The Horizon's Embrace

Life's Ephemeral Beauty: Contemplating the impermanence of life doesn't dim its glow. Instead, it accentuates the beauty of every fleeting moment, urging one to live with authenticity and passion.

Legacy and Lessons: Thoughts often drift to the legacy one leaves behind – the memories shared, the lessons imparted, and the love that will continue to resonate.

Peaceful Acceptance: With the wisdom of years, many find a tranquil acceptance of life's cyclical nature, cherishing the journey and the inevitable horizon that awaits.

As we pause at the edge of these silver shores, gazing at the twilight, it's evident that late adulthood isn't merely an ending. It's a celebration, a culmination of moments, memories, and milestones. With every gentle wave, with every golden hue, there's a story, a lesson, and a reminder to embrace every phase of life with grace and gratitude.

Chapter 10: Embracing the Sunset: Understanding Death and Dying

The Golden Horizon

As day gives way to night, the sun dips below the horizon, casting a warm, golden glow that reminds us of the cyclical nature of existence. Just as every story has a beginning, middle, and end, so does the journey of life. This chapter is a gentle exploration of that final chapter, embracing the sunset with understanding, compassion, and serenity.

The Nature of Life's Final Chapter

The Inevitable Journey: All living beings, from the tiniest blade of grass to the grandest of mountains, have a life cycle. It's a natural progression, from birth to growth and eventually, to a gentle return to the cosmos.

Dying – A Process, Not an Event: Death isn't just a singular moment. It's often a culmination of life's experiences, memories, and lessons. For some, it may be gradual, while for others, it might be sudden, but it remains an integral part of the human experience.

Cultural Lenses: Varied Vistas of Mortality

Global Perspectives: Different cultures have distinct ways of understanding and honoring death. From the vibrant Day of the Dead celebrations in Mexico to the serene Japanese Obon festival, the world offers a myriad of insights into the end of life.

Rituals and Rites: Ceremonies, traditions, and rituals often serve as bridges, helping communities come to terms with the loss while celebrating the departed's life.

Myths and Morals: Tales of afterlife, reincarnation, or ancestral spirits reflect humanity's quest to understand and find meaning in the great beyond.

The Personal Tapestry: Emotions, Experiences, and Enlightenment

Grief's Grasp: The loss of a loved one often brings waves of emotions - sorrow, disbelief, anger, and eventually, acceptance. It's a personal journey, unique to each individual.

Conversations and Closure: Open discussions about end-of-life wishes, memories, or even unresolved matters can provide closure, allowing both the departing and the loved ones to find peace.

Finding Meaning: Many seek solace in spirituality, philosophy, or simply the shared stories of loved ones, drawing strength and understanding from these sources.

Societal Support: Navigating Grief Together

Community and Compassion: In times of loss, communities often rally together, providing emotional support, sharing memories, and offering a shoulder to lean on.

Professional Guidance: Counselors, therapists, and spiritual leaders can offer valuable insights and coping mechanisms, aiding individuals as they navigate the maze of emotions.

Celebrating Life: Memorial services, anniversaries, and other commemorations serve as reminders of the joy, laughter, and love shared, focusing on a life well-lived rather than the loss.

As the sun's glow fades, giving way to a starlit sky, we're reminded that endings are also beginnings. Embracing the sunset isn't just about understanding death but celebrating life, cherishing memories, and fostering connections that, in many ways, transcend the constraints of time. In the grand tapestry of existence, every thread – from the first light of dawn to the gentle embrace of dusk – holds immense value, painting a picture of beauty, resilience, and eternal hope.

Chapter 11: Sailing Through Storms: The Impact of Environment and Genetics

Navigating Uncharted Waters

Imagine setting sail on a vast ocean, equipped with a map and a compass. The map represents our genetic blueprint, handed down through generations. The compass, on the other hand, represents the environment, constantly pointing and guiding us based on external factors. As we journey through life, these two elements constantly intertwine, steering our course and defining our experiences.

Blueprints and Beaches: The DNA Map

The Story Within: Each of us carries a unique genetic code, a story scripted by our ancestors. These genes can dictate our hair color, our risk for certain health conditions, and even some of our behaviors.

Inherited Yet Unique: Even identical twins, with almost the same genetic makeup, can have differences. Why? Because genes provide potentialities, not certainties. Our environment plays a vital role in determining how these potentialities unfold.

The Changing Tides: Environmental Echoes

First Footprints: From the lullabies our parents sang to the schools we attended, early environmental influences leave lasting imprints on our personalities, beliefs, and behaviors.

Relationships and Reactions: Our interactions, friendships, and even brief encounters can shape our perspectives, guiding our responses to future situations.

Landmarks and Life Events: Major life events, be they joyous or challenging, have a profound impact on our character and outlook. A supportive environment can help us navigate these with resilience, while a harsh one might pose challenges.

The Dance of Destiny: Nature and Nurture Interplay

Harmonious Tunes: Our genes might predispose us to a love for music, but it's the exposure to melodies, lessons, and practice that will determine our musical prowess.

Challenges and Choices: Genes might make one susceptible to certain health issues, but lifestyle choices, shaped by environmental factors, play a crucial role in actual outcomes.

Endless Possibilities: While genetics sets the stage, our environment continuously sculpts, refines, and modifies our journey. Each decision, every experience, further tailors the intricate dance of destiny.

Steering the Ship: Can We Influence Our Course?

Embracing Awareness: By understanding both our genetic predispositions and environmental influences, we can make informed decisions, optimizing our well-being and personal growth.

Seeking Favorable Winds: While we can't change our genetic makeup, we can choose environments that nurture our strengths and buffer our vulnerabilities.

Guided Growth: Interventions, education, and self-awareness can help us harness the best of both worlds, steering our ship towards fulfilling destinations.

As we continue our voyage, it becomes clear that the age-old debate of nature vs. nurture isn't about one overshadowing the other. It's about understanding the harmony between the two. Just as a sailor relies on both the map and the compass, we sail through life guided by the interplay of our genes and our environment. And while there may be storms and serene days alike, understanding this dance equips us to navigate life's ocean with grace, resilience, and hope.

Chapter 12: Charting New Courses: Current and Future Research in Human Development

Setting Sail on New Waters

Imagine being an explorer, looking out onto the vast horizon, pondering the mysteries of the uncharted seas. This is the essence of research in human development. Just as every sea wave brings a new story to the shore, every research finding brings a new perspective to our understanding of life's journey.

Current Compass Points: What We Know Now

Brainwaves and Development: With advances in technology, we're now peeking into the human brain like never before. By observing which areas light up during specific tasks, we're learning more about how we think, learn, and feel at different stages of life.

Bonding and Building: Research into human relationships, from family to friends, has illuminated the importance of strong emotional connections for psychological well-being.

Digital Horizons: In today's interconnected world, understanding the impact of technology on human development is vital. From social media interactions to virtual learning environments, we're diving deep into the digital age's effects on growth and learning.

Emerging Waves: What's on the Rise

Nature's Nudge: The concept of 'epigenetics' has recently made waves. Without delving into complex terms, think of it as the environment's ability to nudge our genetic potentials in certain directions.

Mindfulness and Well-being: More and more research is focusing on the benefits of mindfulness, meditation, and holistic approaches to well-being, particularly in stressful modern environments.

Diverse Voices: A growing trend is to include diverse cultural, socio-economic, and individual perspectives. Recognizing that one size doesn't fit all in human development, we're broadening our horizons to capture the human story's full spectrum.

Uncharted Oceans: Future Directions

Interplanetary Perspectives: With talk of Mars colonization and space habitats, future research might explore human development in extraterrestrial environments!

Tech Ties: As Artificial Intelligence and virtual realities become an integral part of life, understanding their long-term impact on cognitive, social, and emotional development will be crucial.

Preserving the Planet: With climate change and environmental concerns, research will inevitably focus on these challenges' psychological and developmental impacts on coming generations.

Just like the boundless ocean, the realm of human development research is vast, deep, and ever-evolving. As we continue to chart new courses, the journey promises to be as exciting as the destination. For, in seeking to understand our paths, we're not just uncovering facts but piecing together the intricate, beautiful mosaic of the human experience.

Chapter 13: A Safe Harbor: Cultivating Well-being Across the Lifespan

Finding Calm Waters in the Journey of Life

Imagine life as a grand, expansive ocean. At times, we might face tumultuous storms or encounter unexpected challenges. In other moments, we might experience smooth sailing under a canopy of stars. Throughout this journey, one thing remains certain: the need for a safe harbor – a place of well-being, peace, and contentment. This chapter will guide you through crafting your own harbor, no matter where you are on life's vast ocean.

Infancy and Early Childhood: Building Strong Foundations

Secure Anchors: Form strong bonds of trust with infants. Responsive and consistent care creates a foundation of security and love.

Playful Explorations: Encourage imaginative play. It's not just fun; it's a child's way of understanding the world and boosting cognitive and emotional growth.

Healthy Horizons: Instilling healthy habits early on, from balanced meals to active playtimes, sets the stage for lifelong physical well-being.

Childhood: Navigating New Waters

Learning Lighthouses: Foster curiosity. Support academic endeavors and always keep the flame of inquisitiveness alive.

Peer Portals: Building friendships is more than just play. It's about learning trust, cooperation, and emotional understanding. Encourage positive social interactions.

Emotional Equilibrium: Teach kids to recognize and express their feelings, ensuring they have the tools to handle life's ups and downs.

Adolescence: Sailing Through Storms

Identity Islands: Adolescents are discovering who they are. Support their quests for identity, understanding that it's okay for them to explore and change.

Guided Compass: While giving them the independence they crave, ensure they have the guidance they need, especially when facing challenging decisions.

Mental Health Maps: Recognize signs of stress or emotional turmoil. Open channels of communication and consider seeking professional help if needed.

Adulthood: Charting Personal Pathways

Balanced Boats: Juggle responsibilities – work, family, personal passions – ensuring none is neglected. It's about harmony.

Relationship Ropes: From partners to friends, relationships keep us anchored. Invest time and emotion in cultivating and maintaining them.

Self-care Shores: Never underestimate the importance of 'me time.' Whether it's a hobby, meditation, or a simple walk, find what recharges you.

Late Adulthood: Reflecting on the Journey

Legacy Lagoons: Encourage seniors to share their stories, wisdom, and experiences. It's therapeutic and enriching for all involved.

Active Atolls: Physical activity remains crucial. Even simple exercises can boost health and mood.

Community Coves: Staying socially active combats feelings of isolation. Community events or even family gatherings can provide much-needed connection.

No matter where you are on life's journey, remember that well-being is a harbor you can always steer towards. It's a blend of the choices we make, the relationships we cultivate, and the moments of introspection we indulge in. And in this vast ocean of existence, knowing that such a sanctuary exists within us can make all the difference.

About Freudian Trips

Welcome to Freudian Trips, your dedicated platform for diving deep into the world of psychology. We are more than just a YouTube channel or a book publisher. We are a beacon of enlightenment, making complex psychological concepts accessible and engaging for all.

Our YouTube channel is a rich repository of psychology made simple. We take the profound and often complex ideas from the world of psychology and break them down into digestible, easy-to-understand content. From the foundational theories of Freud to the cognitive insights of Piaget, we cover a broad spectrum of psychological schools and thoughts, making psychology accessible to everyone, regardless of their background or prior knowledge.

As a book publisher, we take the same approach, transforming intricate psychological theories into comprehensible narratives. Our books are not just collections of words, but vessels of wisdom that make psychology approachable and relatable. We believe that psychology should not be confined to academic circles, but should be available to all who seek to understand the human mind and behavior.

At Freudian Trips, we believe in the power of curiosity and the pursuit of knowledge. We are here to stoke the fires of your curiosity, to guide you on your intellectual journey, and to help you navigate the fascinating world of psychology.

If you are someone who is not afraid to question, to explore, and to learn, then you are in the right place. Join us on this journey of exploration, as we make psychology easy to understand, one concept at a time.

Be sure to visit our Youtube channel at: www.freudiantrips.com/youtube

You can also visit us on the web at www.freudiantrips.com

Welcome to The Freudian Trip community. Stay curious. Stay enlightened.

9 798863 156613